PROBABILITY

MASTERING PERMUTATIONS AND COMBINATIONS

A BEGINNER'S GUIDE

TECHWORLD

© 2017

The information herein is offered for informational purposes solely, and is universal as so. The presentation of the information is without contract or any type of guarantee assurance.

The trademarks that are used are without any consent, and the publication of the trademark is without permission or backing by the trademark owner. All trademarks and brands within this book are for clarifying purposes only and are the owned by the owners themselves, not affiliated with this document.

Introduction

Many tasks in probability theory require that we count the number of ways in which a particular event can happen. For instance, in competitive programming, medium-hard questions require the use of either permutation or combination along with other concepts such as graph theory to solve.

In advanced statistics, permutations and combination are vital in making decisions. For instance, if we know the probability of an event, say a, and the probability of another event say b, we can infer which choice is better. In intelligent gambling, permutations and combinations can be used to determine if you're going to win or not.

In other problem-solving matters, it is often significant to compute the likelihood that a combination of events or an ordered collection of events will happen. Understanding some of the fundamental concepts of probability will provide you with the tools to make informed predictions about events or event combinations. This provides you a solid foundation for understanding probability distributions, the confidence intervals, and hypothesis testing. Permutations and combinations are two essential concepts that you can use to build this foundation.

But, permutations and combinations usually cause a lot of confusion. Specifically, "Which one is which?" and "which one can you use?"

In this book, we delve deeper to provide you with a solid understanding of these fundamental counting theories: permutations and combinations. Specifically, you'll learn the following concepts:

- Basics of permutations and combinations

- Mixing permutations and combinations

- Applications of permutations and combinations in lottery

- Applications of permutations and combinations in poker

CONTENTS

Chapter 1: Overview of Permutations and Combinations

Both permutation and combination involve counting the number of ways that some events can be chosen. Before we delve deeper into discussing permutations and combinations, it is important to understand a general counting principle that allows you to solve different of counting problems such as the problem of counting the number of possible permutations and combination of n objects.

Counting Problems

Consider an experiment that occurs in several phases and is such a manner that the number of outcomes says m at the nth step is independent of the results of the preceding stages. The number m can be different for the various stages. We want to count the number of ways in which the entire experiment can be carried out.

Consider the example below.

You are eating at a City Mall restaurant, and the waiter informs you that you have the following choices for your menu:

a) Two options for appetizers (soup or juice)

b) Three choices for the main course (meat, fish, or vegetable dish).

c) Two choices for dessert (ice cream or cake).

1

How many potential selections do you have for your complete meal? Your menu will be decided in three stages. At each stage, the number of possible selections does not depend on what is selected in the previous phases: two choices at the first stage, three choices at the second, and two options at the third stage.

In particular, you'll have the following choices:

1. Soup, meat and ice cream
2. Soup, meat, and cake
3. Soup, fish and ice cream
4. Soup, fish, and cake
5. Soup, vegetable and ice cream
6. Soup, vegetable, and cake
7. Juice, meat and ice cream
8. Juice, meat, and cake
9. Juice, fish and ice cream
10. Juice, fish, and cake
11. Juice, vegetable and ice cream
12. Juice, vegetable, and cake

Clearly, we have 12 (2 * 3 * 2 = 12) possible menus. You can see that the total number of selections is the product of the number of choices in each phase. In this instance, we have

2 * 3 * 2 = 12 possible menus. The menu example above is an illustration of the following a general counting technique. Next up, let's define a counting technique.

A Counting Technique

The problem is to be carried out in a series of r stages. There are n_1 ways to conduct the first phase; for each of these n_1 ways, there are n_2 ways to carry out the next phase; for each of these n ways, n3 ways to conduct the third step, and so forth. The following product gives the total number of ways in which the entire task can be completed:

$$N = n_1 * n_2 * \ldots * n_r$$

In probability theory, a tree diagram can be used to solve such a counting problem, especially when studying probabilities of events that relates to experiments that take place in phases and for which you are given the probabilities of the outcomes of each phase.

For instance, assuming that the owner of City Mall Restaurant has observed that 80 percent of his clients select the soup for an appetizer and 20 percent select juice. Of those who select the soup, 50 percent preferred meat, 30 percent choose fish as a main course, and 20 percent select the vegetable dish. Of those who opt for juice for an appetizer, 30 percent choose meat meal, 40 percent choose a fish meal, and 30 percent choose the vegetable dish.

We can now use this information to estimate the probabilities at the first two phases as using a tree diagram. For instance, what probability will you assign to the customer who chooses soup and then picks meat as the main course? If 8 of 10 customers selects soup and then 1 out of 2 of these opts for meat, then the proportion 8/10 * 1/2 = 0.4 of the customers choose soup and then meat. This implies the probability distribution for each path through the tree diagram is the product of the probabilities at each of the phases along the tree path.

Now, consider the example below.

We can demonstrate that there are at least two people in Nairobi and Mombasa who have the same three initials. Supposing that each person has three initials, then there will be 26 possibilities for a person's first initial, 26 possibilities for the second initial, and 26 possibilities for the third initial. Therefore, the total of a number of the possible set of initials that we can generate is:

Total numbers of likely initials $= 26^3 = 17{,}576$ possible sets of initials.

This number is smaller than the number of people living in Nairobi and Mombasa. Thus there must be at least two people that have the same three initials.

Let us now consider the often celebrated birthday problem. The celebrated birthday problem is used to demonstrate that any naive intuition can't always be trusted in probability.

The celebrated birthday problem

How many people should we have in a room to make it an auspicious bet (probability of success which is greater than$\frac{1}{2}$) that the two people in the room have the same

4

birthday? Because there are 365 possible birthdays, it is quite tempting to guess that we will require about $\frac{1}{2}$ this number that translates to 183.

With 183 as your best bet, you can surely win this bet. In fact, the total number of people required in the room for a favorable bet is only 23. To show this, we will find the probability Pr that, in a room with r people, there is no duplication of the birthdays for us to have a favorable bet if the likelihood is less than one-half.

Consider the table below

Number of people	Probabilities that all the birthdays are different
20	.5885616
21	.5563117
22	.5243047
23	.4927028
24	.4616557
25	.4313003

Table 1.1: Birthday problem

Supposing that there are 365 possible birthdays for each person (we are ignoring the leap years). We can now order the people starting from 1 to r. For a sample point w, we can now choose a possible sequence of length r of the birthdays each chosen as one of the 365 possible dates.

In general, there will be 365 possibilities for the first component of the sequence, and for each of these selections there will be 365 for the second, and so forth,

5

making 365^r possible series of birthdays. Now, we must determine the number of these sequences that will have no duplication of birthdays. For such a sequence, we will choose any of the 365 days for the first component, then any of remaining 364 for the second person, 363 for the third person, and so forth, until we have to make r selections. For the rth choice, there will be $365 - r + 1$ possibilities. Therefore, the total number of sequences that have no duplications is:

$$365 * 364 * 363 \cdot \ldots * (365 - r + 1)$$

Therefore, supposing that each sequence is equally likely, then we can express it as follows:

$$Pr = \frac{365 * 364 * 363 \bullet \ldots * (365 - r + 1)}{365^r}$$

We can now denote the product as follows:

$$(n)(n - 1) \cdots (n - r + 1)$$

Thus,

$$Pr = \frac{(365)r}{365^r}$$

The celebrated birthday program carries out this computation and prints out the probabilities for the value of r = 20 to 25. Executing this program, you'll get the results shown in Table 1.1. As I had mentioned above, the probability of obtaining no duplication changes from greater than one-half to less than one-half moves from 22 to 23 people.

To see how improbable it is that you would lose your bet for larger numbers of people, you can now run the program again, printing out the values from r = 10 to r = 100 in steps of 10. You'll find out that that in a room of 40 people the odds heavily favor the duplication, and in a room of 100 the odds will be overwhelmingly for a duplication.

Now that you have mastered the concepts underlying counting problems, what next? Next up, we explore the basics of permutations.

Chapter 2: Basics of permutations

Let A be any finite set. A permutation of A is a one-to-one function that maps A onto itself. To specify a specific permutation, we will list the elements of set A and, under them, show where the one-to-one mapping sends each element in the set. For instance, if A = {a, b, c} the possible permutation α would be:

$$\alpha = \begin{matrix} a & b & c \\ b & c & a \end{matrix}$$

By the permutation α, a will be sent to b, b will be assigned to c, and c will be sent to a. The condition that the mapping is a one-to-one implies that no 2 elements of A are sent, by the mapping, into the same element of set A. We can now put the items of the laid down in some order and rename them 1, 2. . . n.

Then, a typical permutation of the set A = {a1, a2, a3, a4} will be written in the form;

$$\alpha = \begin{matrix} 1 & 2 & 3 & 4 \\ 2 & 1 & 4 & 3 \end{matrix}$$

This indicates that a1 was sent to a2, a2 was sent to a1, a3 was sent to a4, and a4 was sent to a3. If you always select the top row to be 1 2 3 4 then, to define the permutation, you need only to give the bottom row, with the understanding that this will tell you where 1 goes, 2 goes, and so on, under the mappings.

When you get done, the permutation will often be called a rearrangement of the n objects 1, 2, 3. . . N.for instance, all the possible permutations, or rearrangements of the numbers A = {1, 2, 3} will be:

123, 132, 213, 231, 312, 321

It is easy to count the total number of possible permutations of the n objects. By our general counting principle, there will be n ways to assign the first element, for each of these we will have n − 1 ways to represent the second object, n − 2 for the third, and so on. Strictly speaking, this proves the following theorem:

The total number of permutations of any set A of n elements is given by the formula:

n * (n − 1) * (n − 2) * . . . ·*1

Obviously, the number that we have just defined in the above formula is the n factorial, and is denoted by n! The expression 0! is equivalent to 1. This makes some formulas to be simpler. The

Sometimes, it is helpful to consider the orderings of subsets of a given set. This will prompt the following definition:

If A is an n-element set, and k is an integer falling between 0 and n, then a k-permutation of set A is an ordered listing of a subset of A of size k that is given by:

n*(n − 1) * (n − 2) * . . . * (n − k + 1)

In a sense, Permutations counts all the possible ways of arranging the elements of a set A. By counting the possible ways; we must be concerned about every last detail, such as the order of each element in set A. Permutations helps to see differently ordered arrangements as the different outputs.

Consider the example below.

Suppose there are five people in a barbecue contest: Andy, Bob, Charlie, David, and Eric. How many different ways can we award the first, the second and the third place ribbons (blue, red and yellow) among the five contestants?

Now, because the order in which the ribbons will be awarded is necessary, should use permutations.

Here's a breakdown:

- Blue Ribbon: There will be five choices: A B C D E assuming that A wins the blue ribbon.

- Red Ribbon: There will be four remaining choices: B C D E assuming that B wins the red ribbon.

- Yellow Ribbon: There will be three remaining options: C D E assuming that C wins the yellow ribbon.

For this example, we have chosen a certain people to win, but this doesn't really matter. All that is important is that we now understand that we have 5 choices at first, then four choices and then three options. The total number of options becomes 5 × 4 × 3 = 60. We had to order the three people out of five. To achieve this, we started with all the five options then take them away one at a time in the sequence four, then three, and so forth until we run out of ribbons.

Since there are 5 choices, we can write the total ways as 5!=5*4*3*2*1. But 120 can only work if we had five ribbons. However, we don't have five; we only have three ribbons. Ideally, we only want 5 × 4 × 3 which is the total number of options. Now, how do we get the factorial to "terminate" at 3?

We must get rid of the 2 × 1. Obviously, 2*1 is 2! This is what will be left over after we choose three winners from the five contestants.

The best and perhaps the easiest way to write this would be 5! / (5 − 3)!

This is just like saying, "use only the first three numbers of 5!"

If we have n items in total and want to choose k of them in a certain order, we will get:

$$\frac{n!}{(n-k)!}$$

And this is the permutation formula that you will always use when computing the number of ways k items will be ordered from n items. We can also express it as:

$$P\ (n,\ k)\ \text{or}\ _nP_k = \frac{n!}{(n-k)!}$$

Let's now turn to some basic examples to help you understand what we have explained in details.

Example 2.1: How many different permutations are there using the letters mid?

Solution

Clearly, the number of letters in the word mid are three. Therefore, the total number of permutations of using the letters mid will be given by:

$$3! = 3*2*1 = 6$$

Assuming that someone randomly mixes the letters of the word mid. What is the probability that the resulting "word" will have i as the middle letter?

Solution: The total number of ways of scrambling the total letters in the word is: 3!

Therefore, to obtain i as the middle for each scrambling, we will need 2! Ways

This gives us $\frac{2!}{3!} = \frac{2}{6} = \frac{1}{3}$

Example 2.2: Consider the word "Tublin." In how many ways can these letters be organized in a row?

Solution

The total number of ways is:

$$6! = 720$$

Suppose that we want the Tub to stick together. How many ways can the letters be organized in a row?

Solution

The total number of ways will be given as:

$$4! = 24$$

Now, assume that the letters in the word "Tublin" are randomly organized in a row. What is the probability that the word "Tub" will sticks together as a unit?

Solution

$$\frac{24}{720} = \frac{1}{30}$$

Example 2.3: 5 midshipmen are lining up to get on the bus so that they go to some exciting, fun mandatory event of a liberty day. How many different arrangements will the Midshipmen line up in?

Solution

The total number of ways $= 5! = 120$

Example 2.4: Consider the problem: MIDN, where MIDN Berrios has 4 midshipmen who work for him: MIDN Murdock, MIDN Ogden, MIDN Vo and MIDN Lees. Now, he has offered to help out and take on the job. We have abandoned all hope of waking up MIDN Miller; there are only three different jobs that need to be done as follows:

- Task 1: Hiding chalks

- Task 2: Finding funny YouTube videos that consume precious class time

- Task 3: Singing Sponge Bob fun song

Assuming that any of the 5 midshipmen can perform any of the 3 jobs, and we want to assign only one midshipman to each task. How many distinct ways can we assign the midshipmen to these three jobs?

Solution

Since we are only interested in 3 jobs and we have 5 men, we will require:

$$\frac{5!}{(5-3)!} = \frac{5!}{2!} = (5)(4)(3) = 60$$

Now that suppose that we have sore throats, and no one can sing the "Sponge Bob fun" song. This leaves us with only two jobs to do:

- Task 1: Hiding chalks

- Task 2: Finding funny YouTube videos that consume precious class time

How many distinct ways can we assign the five midshipmen to these 2 jobs?

Solution

The total number of distinct ways will be equal to:

$$\frac{5!}{(5-2)!} = \frac{5!}{3!} = (5)(4) = 20$$

14

Now, let us modify the problem again. Suppose the YouTube site has gone down. We'll now be left with one job to do:

Task 1: Hiding chalk

How many distinct ways can we assign the five midshipmen to this single job?

Solution

$$\frac{5!}{(5-1)!} = \frac{5!}{4!} = 5$$

Example 2.5: How many two letter words can be generated by choosing two different letters from the alphabet?

Solution

$$P(26,2) = \frac{26!}{(26-2)!} = (26)(25) = 650$$

Example 2.6: How many different permutations exist for the seven letters in the word "SCHMIDT" taken four letters at any given time?

Solution

$$P(7,4) = \frac{7!}{(7-4)!} = (7)(6)(5)(4) = 840$$

Example 2.7: All the Midshipmen have entered a raffle by placing their names in the hat. Now, a name is drawn from the hat at random for first prize; then another name is drawn from the hat for a second prize. How are many distinct winning name selections possible?

Solution

$$P(12,2) = \frac{12!}{(12-2)!} = (12)(11) = 132$$

Example 2.8: Joseph Cramer is organizing a string of Christmas lights in his room. He has three red, four orange and two green little Christmas light bulbs. How many different lighting arrangements can he make with for a string of nine Christmas lights?

Solution:

$$\frac{9!}{3!4!2!} = 1260$$

Example 2.9: 11 people are traveling on a trip to Washington DC in three cars. The cars hold 2, 4 and 5 passengers respectively. In how many ways can these people transport themselves to Washington DC?

Solution

$$\frac{11!}{2!4!5!}$$

Chapter 3: Basics of Combinations

The combination is a technical term that means "selections." We use the term combinations to refer to the number of different sets of a certain input size that can be chosen from a larger collection of objects where order doesn't matter.

Let's consider the example below.

Suppose you are given 7 different points on a Cartesian plane in such way that there no 3 of them are on the same straight line. How many different straight lines that pass through 2 of the given points?

For any 2 distinct points in the Cartesian plane there is one straight line that passes through these points. In particular, every pair of the given 7 points will determine the straight line that passes through these points.

Conversely, any straight line that passes through the pair of the given 7 points will determine this pair uniquely since no three of the 7 points are on the same straight line. This implies that the number of all the straight lines that pass through two of the given points is equal to the number of all different pairs composed of these 7 points.

The order of the points inside the Cartesian pair doesn't matter while counting the pairs since the permutations of the points inside the pair provides the same straight line. Now you can easily compute the number of pairs. You can choose any of 7 points as the first point in the pair. This gives you 7 independent selections.

Then you can choose any of 6 left points as the second point in the pair. This provides you 6 independent choices. In this manner you count each pair twice since permutation of the points inside the pair will remain the same. Therefore, the number of different pairs composed of the given 7 points is equal to:

7*6/2=21

Therefore, we have found out that the number of straight lines that pass through the 2 of the given 7 points is equal to the number of all the different pairs composed of the given 7 points. In turn, this number is equivalent to:

7*6/2=21

Let us consider another example.

Assuming that you have 7 points in on Cartesian plane such that neither three of them are on the same straight line nor 4 of them are in the same circle. How many different will you obtain if each is passing through three of the given points?

First, for any 3 points on the plane that aren't on the same straight line, there is always a circle that passes through them. This circle is circumferential to the triangle built on the 3 points as its vertices. This circle is uniquely specified for any three points on the Cartesian plane that are not in a same straight line. Therefore, any triple of the 7 given points will have the unique circle passing through these three points.

Now we can easily compute the number of triples. We begin by selecting any of the 7 points as the first point of the triple. This will give us 7 independent choices. Next, we select any of the 6 remaining points as the second point of the triple. This will give us 6 independent choices. Finally, we select any of the 5 remaining points as the third point of the triple.

This way, you will count each triple 2*3 = 6 times according to the number of permutations of the points inside the triple. Therefore, the number of different triples composed of the 7 points is equal to:

$$7 * 6 * 5/2 * 3 = 35$$

Both the first and the second example are about counting the number of certain subsets of the given finite set of elements.

Here is how we can define combinations:

Assuming that there is a set of n distinct objects. A combination is a selection of the m elements of the given set, where m < n, without considering the order in the outcome subset. In other words, we say it is a combination of n things that are taken m at a given time.

The total number of combinations of n things that are taken m at a time is denoted by:

$$^mC_n$$

Ideally, a combination of n things taken m at a given time is a selected subset of m elements of the original set of n elements. Consequently, the number of combinations of n items taken m at a given time is the number of subsets of m elements of the original set of n elements.

I know you're now asking, "Why is the definition focusing on distinguishable objects if we aren't considering the order of items in the outcome subset?"

As a matter fact, we don't require to have the distinguishable objects to make the selection. However, we need to have the distinct objects if we were to have distinct different combinations while counting the total number of the various combinations of n items taken m at a given time.

Mathematically, we can express combination as follows:

$$^mC_n = \frac{n!}{m!(n-m)!}$$

Consider the following examples:

Example 3.1: 4 students are to be chosen from 18 students of class to represent the class in the school debate. In how many ways can the 4 students be selected from among the body of the class?

Solution

Clearly, the selection of the 4 students from the body of 18 students in the class is a combination of 18 students taken 4 at a given time.

Therefore, the total number of such combinations is:

$$^{18}C_4 = \frac{18!}{4\,!(18-4)!} = 73440/24 = 3060$$

Example 3.2: The Quality Assurance Service of a given organization has to test a sample of 8 tires selected from among the 100 tires. In how many ways can the 8 tires can be chosen for testing from the set of 100 tires?

Solution

Selecting a subset of 8 different tires from the set of 100 tires is a combination of 100 tires taken 8 at a given time.

The total number of such combinations is:

$$^{100}C_8 = \frac{100!}{100\,!(100-8)!} = 2000945100$$

Example 3.3: In how many ways a subset of 7 different domino bones is chosen from a standard set of 28 tiles?

Solution

Obviously, the order of the domino bones in the chosen subset doesn't matter.

Selecting a subset of 7 different domino bones from the standard set of 28 tiles is a combination of 28 tiles taken 7 at a time.

The total number of such combinations is:

$$^{28}C_7 = \frac{28!}{7\,!(28-7)!} = 1184040$$

Chapter 4: Mixed Permutations and Combination

We have so far explored permutation and combination as some of the counting techniques. In particular, we've learned the following counting principles:

If you can break a problem into r steps, with m_1 ways of performing step 1, m_2 means of performing step 2 (no matter what you do in step 1) and so forth up to m_r ways of performing step r (no matter what you do in the preceding steps), then the total number of ways in which you can complete the task is:

$$m_1 {}^* m_2 {}^* ... {}^* m_r$$

- The total number of arrangements of n objects taken k at a time is called permutation. A permutation can formally be defined as:

$$P(n, k) = \frac{n!}{(n-k)!}$$

- The total number of distinct ways of choosing a subset of m objects from a set that has n objects, where order doesn't matter is called combination. The combination can formally be defined as follows:

$$C(n, m) = {}^m C_n = \frac{n!}{m!(n-m)!}$$

The above counting techniques can help you solve any problem that you may encounter. However, not all problems can be solved by using only one method. Other tasks will require you to apply more than one method.

When asked to count the total number of objects in a set, it is important to think of how you may complete the task of constructing the object in the set. It also helps to keep the technique of overcounting in mind. So, how do you decide to select the appropriate method?

Here are considerations that you should make before settling on one method:

- If the task involves making two or more separate decisions and asks you about the combined number of probabilities, then you can use a Multiplication Principle.

- If the problem involves a group of objects and asks you a question about how many ways they can be chosen or assigned or put in order, use the flowchart below to guide you:

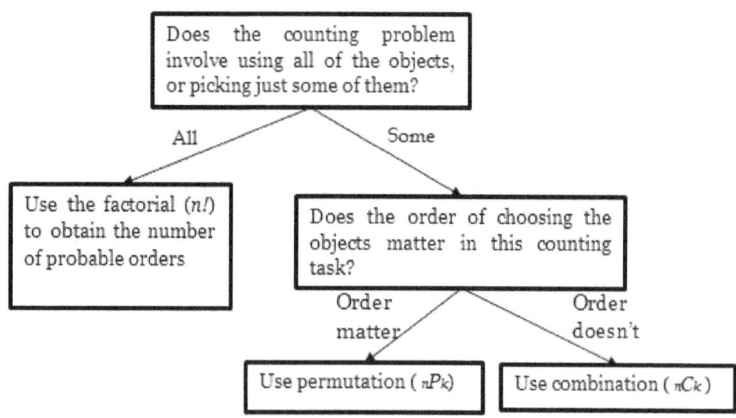

Let us now consider the following problems.

Example 4.1: An experiment consisting of rolling a 20-sided die three times. The number on top of each die is then recorded. The numbers displayed on top of each die are written down in the order in which they are observed. How many possible ordered triples of the numbers can result from the experiment?

Solution

In this example, the triple set (17,10, 3) isn't the same result as the triple set (3, 10,17).

There are 20 ways in which each throw can come up, and the order is vital, so the answer is:

$$20*20*20 = 20^3 = 8000.$$

Example 4.2: When you purchase a Powerball ticket, you choose 5 different white numbers from among the numbers 1 through 59 (the order of selection doesn't matter), and one red number from among the numbers starting from 1 to 35. How many different Powerball tickets can you purchase?

Solution

You should select 5 distinct white numbers, so you can do this in:

$$C(59,5) = 5006386 \text{ ways}$$

To select the red number, you need:

$$C(35, 1) = 35 \text{ ways}$$

Thus, the total number of tickets is:

$$C(59, 5) *P(35, 1) = 5006386*35 = 175223510 \text{ ways}$$

Example 4.3: A bag contains 15 marbles. Out of these 15 marbles, 10 are red while 5 are white. Now, 4 marbles are chosen from the bag.

Before you begin to figure out how you'll count, you should understand the ambiguity in this problem. Suppose on the first draw, you select four red marbles, and on the second draw, you choose different 4 marbles, are they considered the same sample or not?

In this problem, we'll assume that they are not the same sample. For instance, you could imagine that the marbles are numbered with each having a different number so that you can distinguish marbles of the same color. This way of thinking will be very helpful for calculating probabilities later.

Let us now answer some questions regarding the bag containing 15 marbles.

a) How many are (different) samples of size 4 possible?

Solution

The order of counting doesn't matter, but the numbers do so we are choosing the 4 elements from a set of (10 + 5) elements. Therefore, the number of different samples of size 4 are:

$$C (15, 4) = 1365$$

b) How many distinct samples of size 4 consist entirely of the red marbles?

Solution

The order doesn't matter, but the numbers are so we are choosing 4 elements from the set of 10 elements. Therefore the total number of a sample size of red marbles is:

$$C\,(10,\,4) = 210$$

c) What the total number of samples that have 2 red and 2 white marbles?

Solution

We can choose 2 numbered red marbles in $C\,(10,\,2)$ ways and the 2 numbered white marbles in $C\,(5,\,2)$ ways. Neither selection will affect the other, so our answer is:

$$C\,(10,\,2) * C\,(5,\,2) = 45\,*10 = 450 \text{ ways}$$

d) How many different samples of size 4 have exactly 3 red marbles?

Solution

We can choose 3 numbered red marbles in $C\,(10,\,3)$ ways and another 1 numbered white marble in $C\,(5,\,1)$ ways. Neither selection will affect the other, so the total number of different samples of size 4 that have exactly 3 red marbles becomes:

$$C(10, 3) * C(5, 1) = 120 * 5 = 600$$

e) How many samples of size 4 have at least 3 red marbles?

Solution

The answer to this question is the number of samples with the 3 red marbles plus the number of samples with 4 red marbles. We can choose 4 numbered red marbles in C (10, 4) ways and the 0 numbered white marbles in C (5, 0) ways. Neither choices will affect the answer, so the total number of samples of a size that have at least 3 red will be:

$$C(10, 3) * C(5, 1) = 210 * 1 = 210.$$

Now, from our last example, there are 600 ways to choose samples with exactly 3 red marbles. Therefore, our answer is:

$$600 + 210 = 810$$

f) How many different samples of size 4 contain at least one red marble?

Solution

We can figure out the answer as: "the number with exactly 1" plus "the number with exactly 2"+ . . . "the number with exactly 4". This is equivalent to:

$$C(10,2)*C(5,3)+C(10, 3)*C(5,2)+C(10, 3)*C(5,1)+C(10; 4)*C(5; 0)$$

This is equal to:

$$10*10+45*10+120*5+210*1 \ = \ 100+450+600+210 \ = \ 1360$$

Example 4.4: In this example, we will use standard deck cards to demonstrate mixed permutations and combinations. Remember that any standard deck of cards has 52 cards. The cards can be grouped according to suits or denominations. In any standard deck card, there are 4 suits (hearts, diamonds, spades and clubs). Each suit has 13 cards in total. There are 13 denominations of Aces, Kings, Queens... Twos with 4 cards in each denomination. A poker hand has a sample of size 5 drawn from the deck.

a) How many poker hands have 2 Aces and 3 Kings?

Solution

You can choose Aces in C (4, 2) ways and the Kings in C (4, 3) ways. Neither of the choices will affect your answer. Therefore, the total number of poker hands that have 2 Aces and 3 Kings becomes:

$$C \ (4, 3) \ * \ C \ (4, 2) = 6 \ *4 = 24$$

b) How many poker hands have 2 Aces, 2 Kings and a card of a different denomination?

Solution

You can choose 2 Aces, 2 Kings in C (4, 2)* C (4, 2) = 6 * 6 = 36 ways. You can choose the remaining card in any of 52 * 8 = 44 ways. Therefore, the final answer is:

$$36 * 44 = 1584$$

c) How many distinct Poker hands have 3 cards from 1 denomination and 2 from another that forms a full house?

Solution

There are 13 different ways to select the first denomination. Then are then C (4, 3) ways to select 3 cards of the denomination. There are 12 ways to pick the second denomination and thereafter C (4, 2) ways to select 2 cards of that denomination.

Therefore, we will have:

$$13* C (4, 2) * 12 * C (4, 3) = 13*4*12*6 = 3744 \text{ ways}$$

Example 4.5: The Mombasa Village club has 20 members. Out of 20 members, 5 are seniors, 4 are juniors, 2 are sophomores, and 9 are freshmen.

a) In how many ways a club be chosen for a president, a treasurer, and a secretary if every member of that club is eligible for each position and no member can hold two positions at the same time?

Solution

There are 20 members in total, so selecting 3 members from 20 members will yield P (20, 3) ways. Note that we are now selecting an ordered subset of 3 distinct elements from a set of 20 members.

b) In how many ways can a club be picked from a group of 5 members that should attend the AU meeting in Addis Ababa?

Solution

The answer is C (20, 5) ways. Note that this time we need a subset of all the members that has 5 elements, but the order is not important.

c) In how many ways can a club be selected from a group of 5 members to attend the next meeting in AU if all the members of the group must be freshmen?

Solution

The answer is C(9,5) because you now must choose your subset from a set of 9 freshmen.

d) In how many means can a group of 5 be selected if there must be at least 1 member of each class?

Solution

There are 5 ways to choose a senior, 4 ways to choose a junior, 2 ways to choose a sophomore and 9 ways to pick a freshman. This gives $5*4*2*9 = 360$ ways to pick a subset with the 4 elements containing one member of each class. When you do this, there will be $20*4 = 16$ members left, and you may select any one of these steps to round out the group. Therefore, the answer is $360*16/2 = 2880$ ways. We are dividing by 2 because each set of 5 elements selected by this procedure must occur twice.

31

Chapter 5: Applications of Permutations and Combinations in Lottery

The jackpot payoff for the lottery continues to increase as the new Mega Millions and the

Powerball lottery games gain recognition. Now more than ever, an individual can win millions of dollars by correctly choosing a lineup of numbers from a set of roughly 50 numbers. The overall chances of prizewinning are subtle. However, many people are willing to risk the $1 ticket price to earn a shot at the grand prize.

This chapter explores the counting principles of winning money from playing the lottery to identify if the trends or patterns exist in the lottery results. For instance, how do we apply the counting principle to find out the chances of winning simple number lottery games? And more specifically, how do we apply different strategies to design a lottery game? Let's recap what we have learned so far.

Counting Principles

As we have learned in the previous sections, the counting principle determines the number of possible outcomes for a sequence of any independent events. The term "independent" is used to show that the events shouldn't depend on the order in which they are finished. For instance, let's say that you have two events,

E1 and E2, where E1 can occur in n1 ways, and E2 can happen in n2 different ways.

To find the total number of ways for the two events to occur, we apply the counting principle: $n_1*n_2*...*n_k$ where k is the total number of possible outcomes.

Permutations
The total number of arrangements of n objects taken k at a time is called permutation. A permutation can formally be defined as:

$$P(n, k) = \frac{n!}{(n-k)!}$$

Combinations
The total number of ways of selecting a subset of m objects from a set that has n objects, where order doesn't matter is called combination. The combination can formally be defined as follows:

$$C(n, m) = {}^mC_n = \frac{n!}{m!(n-m)!}$$

Now that we have our basics, what next?

Let's dive in to explore practical examples of applications of permutations and combinations in the lottery.

Example 5.1: Sportpesa Lottery provides several variations of lottery games. In one of its simplest games called Big 4, a player selects any four-digit number and places a bet that ranges from $0.50 to $5.00. The player will win if his/her number is randomly in the daily drawing. The payoffs for the lottery games are determined by one's odds of winning.

The odds that are for an event are usually expressed as a ratio of the likelihood that the event will happen in the probability that an event won't happen. Here are 5 scenarios that are likely to arise:

a) Play It Straight where the player plays 4 different digits and wins only with an exact match.

b) Play it boxed where the player plays 3 of the same number and 1 other number. The player wins if the digits is drawn in any order.

c) Box 2 pairs where the player chooses 2 pairs of numbers. The player wins if the digit is drawn in any random order.

d) Box 1 pair plus 2 digits where the player chooses 1 pair of a number and 2 other digits. The player wins if the number is drawn in any random order.

e) Box 4 different digits where the player plays 4 different numbers. The player wins if the number is drawn in any random order.

Solutions

To compute the odds of winning, you must know the total number of distinct 4-digit numbers. Now, there will be 10 ways to choose the first number, 10 ways to choose the second number, 10 ways to choose the third number, and 10 ways to select the fourth number.

Therefore, the total number of different 4-digit numbers that one can select is:

$$10*10*10*10 = 10000$$

Let us now compute the odds of winning for 5 scenarios that we described above.

#1: Scenario 1: Play it straight

There is exactly one chance to win (1/10,000). This is because the order of numbers matters in computations.

#2: Scenario 2: Play it boxed

Because the number can be drawn in any order, there will be 4 chances to win (4/10000). These numbers can be in any of the following orders such as abbb, babb, bbba, or bbab.

#3: Scenario 3: Box 2 Pairs

Because the number can be drawn in any order, there will be 6 chances to win (6/10000). These numbers can be in any of the following orders: aabb, abab, bbaa, baba, abba, or baab.

#4: Scenario 4: Box 1 Pair + 2 Digits

Because the number can be drawn in any order, there will be 12 chances to win (12/10000). These numbers can be in any of the following orders: aabc, aacb, cbaa, abca, bcaa, baac, caab, abac, acba, acab, baca, and caba.

#5: Scenario 5: Box 4 Different Digits

Because the number can be drawn in any order, there will be 24 chances to win (24/10000). The order in which the 4 numbers are arranged doesn't matter, so the total number of permutations becomes 4 * 3 * 2 * 1 = 24

Expected value of a lottery

The Expected value for a probability distribution is the most likely value of a random variable. For instance, if you're dealing with an investment decision, then the Expected value is the average value of all possible payouts. To compute the Expected value of a random variable, you must first multiply each possible payout by its probability of occurring, followed by summing all of the products together.

Consider the following examples.

Example 5.2: Let's say that you are rolling a 6-sided die. If you roll a 3, then you automatically win $5.00. On the other hand, if you don't roll a 3, then automatically pay $1.00. What is the expected value of this game?

Solution

You should note that since the probability of rolling a 3 is 1/6, the probability of not rolling a 3 becomes 5/6. Thus, the Expected value becomes:

P (3) * (5) + P (not 3)*(-1) = (1/6) * (5) + (5/6) * (-1) = 5/6 - 5/6 = 0.

In this case, we'll say that the game is far since the expected value is 0.

Example 5.3: *In a certain lottery system, a player selects three digits, which must be in a particular order. Note that the numbers can lead with the digit 0, so the numbers such as 056 or 009 are acceptable. Digits may also be repeated. In each drawing, a 3-digit sequence is selected. Any player that picks matching for all three numbers in the correct order automatically receives a payout of $500.*

Compute the following:

a) The probability of winning this lottery game.

b) The Expected value of winning if it will require $3 to play 1 game.

c) The fair cost for a person to play this lottery game.

Solution

a) The probability of winning the lottery game

Suppose that we select the numbers 2, 9, and 2, in that order. On the first draw, the probability of drawing a 2 out of the 10 possible numbers is 1/10. Since the digits may be repeated, the 2 then is replaced. On the second draw, the probability of now getting a 9 is still 1/10. The 9 is then replaced. So, on the third draw, the probability of getting a 2 is still 1/10.

To compute the probability of winning the game, we apply the multiplicative principle of counting. Therefore, the overall probability of winning is:

$$1/10 * 1/10 * 1/10 = 1/1000$$

b) The Expected value of winning

The expected value of winning is:

$$P \text{ (winning)} * \$497 + P \text{ (not winning)} * \$-3 = (1/1000) * (497) + (999/1000) * (-3) = 0.497 - 2.997 = -2.5$$

What does the answer imply? A negative outcome means that you're expected to lose the game.

c) The fair cost for playing the game

The expected value for any fair game should be 0. In this game, the probability that you'll win is still 1/1000.

We can set the equation for the Expected value to be equal to 0 to solve for the fair cost of the game. Therefore, we'll have the following equation:

$$(1/1000) * (500 - x) + (999/1000) * (-x) = 0$$

This expression will simplify to:

$$0.5 - 0.001x - 0.999x = 0$$

When simplified further, it becomes:

$$0.5 - x = 0$$

The above expression when solved produces a value of 0.50. This implies that the game should cost 50 cents to play for it to be considered fair game.

Example 5.4: *Mega Betin is a national lottery game that is well-known for its large rollover jackpot payoffs. In this game, each player selects 6 total numbers. Out of the 6 numbers, 5 different numbers that range from 1 to 56 are considered the "white numbers, " and 1 number from 1 to 46 is considered the "Mega number." A player can only win the jackpot if he/she matches all the 6 numbers selected in a drawing.*

Compute the following:

a) The probability of winning the Mega Betin jackpot.

b) The probabilities of the other winning combinations.

c) The Expected value of the game to find out whether it is worth purchasing a $1 lottery ticket for an opportunity to win the jackpot prize. Use the jackpot value of $42 million for your computation.

Solutions

a) The probability of winning the Mega Betin jackpot

The numbers of ways to choose 5 numbers from a range of 56 numbers may be computed using combinations. Therefore, this becomes:

C (56, 5) = 3819816 ways

The total number of ways to choose 1 number from a range of 46 numbers is 46. Therefore, the total number of Mega Betin combinations becomes:

3819816 x 46 = 175711536 ways

Now, there is only one way that the first 5 numbers on the lottery ticket can match the 5 selected white numbers. There is also only one way for the 6th number on the lottery ticket to match the Mega Betin Number. This means that there is one way to win the jackpot!

Therefore, the probability of winning the jackpot is 1/175711536

b) The probabilities of other winning combinations

The probabilities of other winning combinations are

i. *Match all the 5 white numbers but not the Mega Betin number (payoff =$250000).* There is only one way in which the first 5 numbers on your lottery ticket can match the

40

5 selected white numbers. Therefore, the number of ways becomes C (5, 5). There are 45 ways in which your 6th number will match any of the 45 losing Mega Betin numbers. This becomes C (45, 1).Therefore, the total number of ways to achieve this combination would be 1 * 45 = 45 which leads to the probability of 45/175711536. When simplified further, it becomes to about "1 chance in 3904701."

ii.   *Match 4 out of the 5 white numbers and the Mega Betin number (payoff = $10000).* There are 5 ways that 4 of the first 5 numbers on the lottery ticket can match the 5 selected white numbers C (5, 4). There are 51 ways for the 5th white number to match any of the 51 losing white numbers C (51, 1). There is one way for your 6th number to match the winning Mega Betin number C (1, 1). The total number of ways to achieve this combination would be 5 * 51 * 1 = 255 which leads to a probability of 255/175711536. When simplified further, it becomes "1 chance in 689065."

iii. *Match four out of five white numbers but not the Mega number (payoff =$150).* There are 5 ways that 4 of the first 5 numbers on the lottery ticket can match the 5 chosen white numbers C (5, 4). There are 51 ways for your 5th white number to match any of the 51 losing white numbers C (51, 1). There are 45 ways for your 6th number to match any of the 45 losing Mega Betin numbers C (45, 1). The total number of ways to achieve this combination would be 5 * 51 * 45 = 11475 which leads to a probability of 11475/175711536. When simplified further, it becomes "1 chance in 15313."

iv. *Match 3 out of the 5 white numbers and the Mega Betin number (Payout = $150).* There are 10 ways that 3 of the first 5 numbers on the lottery ticket can match the 5 selected white numbers C (5, 3). There are 1275 ways for 2 of white numbers to match any of the 51 losing white numbers C (51, 2). There is only one way for your 6th number to match the winning Mega Betin number C (1, 1). The total number of ways to achieve this combination would be 10 * 1,275 * 1 = 12750, which leads to a probability of 12750/175711536. This will simplify to about "1 chance in 13781."

v. *Match 3 out of the 5 white numbers but not the Mega Betin number (payoff =$7).* There are 10 ways that 3 of the first 5 numbers on the lottery ticket can match the 5 selected white numbers C (5, 3). There are 1275 ways for 2 of the white numbers to match any of the 51 losing white numbers C (51, 2). There are 45 ways for your 6th number to match any of the 45 losing Mega numbers C (45, 1). The total number of ways to achieve this combination would be 10 * 1,275 * 45 = 573750, which leads to a probability of

573750/175711536. This will simplify to about "1 chance in 306."

vi. *Match 2 out of the 5 white numbers and the Mega number (payoff =$10).* There are 10 ways that 2 of the first 5 numbers on your lottery ticket can match the 5 selected white numbers C (5, 2). There are 20825 ways for 3 of the white numbers to match any of the 51 losing white numbers C (51, 3). There is one way for your 6th number to match the winning Mega number C (1, 1). The total number of ways to achieve this combination becomes 10 * 20,825 * 1 = 208 250, which leads to a probability of 208250/175711536. This will simplify to about "1 chance in 844."

vii. *Match 1 out of the 5 white numbers and the Mega number (payoff = $3).* There are 5 ways that 1 of the first 5 numbers on the lottery ticket can match the 5 selected white numbers C (5, 1). There are 249900 ways for 3 of white numbers to match any of 51 losing white numbers C (51, 4). There is only 1 way for your 6th number to match the winning Mega number C (1, 1). The total number of ways to achieve this combination becomes 5 * 249900 * 1 = 1249500, which leads to a probability of 1249500/175711536. This will simplify to about "1 chance in 141."

viii. *Match 0 out of the 5 white numbers and the Mega number (payoff = $2).*There is only one way that none of the first 5 numbers on the lottery ticket can match the 5 selected white numbers C (5, 0). There are 2349060 ways for 5 of the white numbers to match any of 51 losing white numbers C (51, 5). There is only one way for your 6th number to match the winning Mega Betin number C (1, 1). The total number of ways to achieve this combination would be 1 * 2349060 * 1 = 2349060, which leads to a probability of 2349060/175711536. This will simplify to about "1 chance in 75."

c) The Expected value of winning

The Expected value of winning is P (Jackpot) * $42000000 + P (match 5 white, not Mega) * $250,000 + P (match 4 white and Mega) * $10,000 + P (match 4 white, not Mega) * $150 + P (match 3 white and Mega) * $150 + P (match 3 white, not Mega) * $7 + P(match 2 white and Mega) * $10 + P(match 1 white and Mega) * $3 + P(match 0 white and Mega) * $2 P(not winning) * $-1 = -0.55

You are expected to lose in this game!

Chapter 6: Permutations for Poker Games

Poker is one of the several games that involve the use of a 52-card deck of playing cards. The 52 cards are usually categorized by 13 ranks from Two through Ace (Aces may be counted as both higher than the King and lower than the Two when needed, however, will be counted as one at a time in a hand), and by the four suits: hearts, diamonds, spades, and clubs. In the game of poker, players will attempt to assemble the best 5-card hand depending on the definitions of each hand that you'll make.

10 hands can be made during the game:

i.   Royal Flush. All the 5 cards will be of similar suit and are of the sequence "10 – J – Q – K – A."

ii.   Straight Flush. All the 5 cards will be of the same suit and are sequential in the rank (note that the royal flush is the highest-ranked straight flush)

iii. Four-of-a-Kind that is usually abbreviated 4OAK. A hand where the four cards are all of the same ranks.

iv.   Full House. A hand has one pair and a 3-of-a-kind of a different rank than the pair.

v. Flush. All the 5 cards will be of the same suit but not all sequential in rank.

vi. Straight. All the 5 cards are sequential in rank but aren't of the same suit.

vii. Three-of-a-Kind that is usually abbreviated as 3OAK. A hand where 3 cards are all of the same rank and the other 2 are each of distinct ranks from the 3OAK and each other.

viii. Two Pair. 2 pairs of 2 cards of the same rank (where the ranks of each pair are different in rank to avoid similarity with 4OAK.

ix. One Pair. Only 2 cards of the 5 are of the same rank with the other 3 cards all having the different ranks from each other and from that of the given pair.

x. High Card. The hand in which no better hand was made.

Poker games have many forms, some of which will be investigated in this chapter. One such variation is the "stud" poker in which the player must hold all the cards that he/she is given. This is different from the "draw" poker in which the player can draw any number of replacement cards after being dealt the first 5 in the attempt to improve his/her hand. Other forms include the use of jokers and wild cards.

To solve poker games, we will still apply the counting principles (multiplicative, permutations and combinations) that we have learned so far.

Consider the following examples.

Example 6.1: In 5-card stud game, each player is dealt 5 cards to make the best 5-card hand possible. Ans since there are 52 cards in the deck, there will be C (52, 5) = 2598960 possible combinations of the 5 card hands possible. We will evaluate the numbers of hands in the typical order of the rank of each hand, beginning with straight flushes (because a royal flush is simply the highest-ranked straight flush we will include it in the discourse of straight flushes, but provide it no additional significance).

a) *Straight Flush*

To have a straight flush the hand should consist of all the 5 cards being of the similar suit and all in numerical order. There are 10 possible sequences (A – 5, 2 – 6... 9 – K, and 10 – A). Because there are 4 suits, then the total number of straight flushes that are possible becomes 10 * 4 = 40, with the highest 4 (each a straight flush "10 – A" of one of the four suits) being the royal flushes.

b) *Four-of-a-Kind (4OAK)*

To have a 4OAK the hand should have all of the cards of 1 of the 13 available ranks plus one other card. Now, it doesn't matter what the last card is. There is only C (4, 4) = 1 combination of all the 4 cards of one rank, and there will be 48 cards left to select from after the 4OAK is obtained. Thus, there are 13 * 4C4 * 48 = 624 possible 4OAK.

c) Full House

Because a full house has the variation of one pair and a 3-of-a-kind, then there will be 13

* 12 = 78 choices for the given ranks of the pair and the 3OAK (note that we don't have to remove permutations from the selections since there is a difference in which of the pair or the 3OAK gets which rank. For instance, a full house has two 4s, and 3 9s is different than one with two 9s and three 4s. There are C (4, 2) = 6 choices for the pair in its rank and C (4, 3) = 4 choices for the 3OAK. Therefore, there will be 12 * 13 * C (4, 2) * C (4, 3) = 3744 possible full houses.

d) Flush

A hand that is a flush should have all 5 cards being of the same suit. Each of the 4 suits will have C (13, 5) = 1287 possible 5-card hands that are all of the same suit. However, some of these combinations are also the straight flushes. Therefore, the 40 straight flushes should be removed from the count. This becomes 4 * C (13, 5) − 40 = 5108 possible flushes.

e) Straight

A hand that is straight should have 5 cards sequential in rank but with all 5 not all of the same suit. Using the similar arguments from the straight flushes and flushes, we will have 10 sequences where there are 4 distinct choices for the particular card in each rank. Therefore, there are 4^5 = 1024 possible ways to select the cards in each sequence. Eliminating the 40 straight flushes will result in the number of straights being equal 10 * 45 − 40 = 10,200.

f) Three-of-a-Kind (3OAK)

This hand should have 3 cards being of the same rank with the other 2 cards not improving the hand. There are 13 ranks to select from for the 3OAK and C(4,3) combinations of 3OAKs within each rank. There are (48 * 44)/2 possible combinations for the last 2 cards (here we have to divide by 2, which is just 2! to eliminate the permutations that would double our count. In any poker, the order in which the cards appear doesn't matter. Therefore, there will be 13 * C (4, 3) * (48 * 44)/2 = 54912 possible 3OAKs.

g) 2 Pair

There are C (13, 2) combinations to select the 2 ranks for the 2 pair and C (4, 2) ways to select the pair in each rank. There are 44 cards possible for the 5th card so as not to improve the hand. Therefore there will be C (13, 2) * C (4, 2)^2 * 44 = 123552 possible 2 pair hands.

h) 1 Pair

Using similar arguments for the previous hands there will be 13 ranks to select from for the pair and C (4, 2) possible pairs per rank, and (48 * 44 * 40)/6 possible ways to choose the other 3 cards (again we want to eliminate permutations and keep only the combinations so we must divide by 3! the number of permutations of the 3 cards). This leaves us with 13 * C (4, 2) * (48 * 44 * 40)/6 = 1098240 possible one pair hands.

i) High Card

The set of high-card hands is a complement to the set of all the other hands. This means that the totalll number of high card hands becomes 2598960 − 40 − 624 − 3744 − 5108 − 10200 − 54912 − 123552 − 1098240 = 1302540.

Resources and Further Readings

Below is a list of websites for useful Permutations and Combination resources:

i. http://www.meteor.iastate.edu/~jdduda/portfoli o/492.pdf

ii. http://webpages.uncc.edu/ras/Logic/chapt43.pd f

iii. https://www.dartmouth.edu/~chance/teaching_ aids/books_articles/probability_book/amsbook. mac.pdf

iv. http://www.statisticshowto.com/how-to-solve-permutations-and-combinations-problems/

v. http://www.tip.sas.upenn.edu/curriculum/units/ 2011/02/11.02.07.pdf

vi. http://www.uhigh.ilstu.edu/math/thompson/Pre calc/Probability%20and%20combinations/combi nations%20and%20permutations.pdf

vii. http://infolab.stanford.edu/~ullman/focs/ch04. pdf

viii. http://www.unco.edu/nhs/mathsci/facstaff/robe rson/CourseDocs/MATH%20182/Activities/Com binations%20and%20Permutations.pdf

ix. http://www.stat.wisc.edu/~ifischer/Intro_Stat/L ecture_Notes/APPENDIX/A1._Basic_Reviews/A 1.2_-_Perms_and_Combos.pdf